Cheaper, Faster, Better

RECENT TECHNOLOGICAL INNOVATIONS

by Cynthia Swain

Editorial Offices: Glenview, Illinois • Parsippany, New Jersey • New York, New York
Sales Offices: Needham, Massachusetts t• Duluth, Georgia • Glenview, Illinois
Coppell, Texas • Ontario, California • Mesa, Arizona

Every effort has been made to secure permission and provide appropriate credit for photographic material. The publisher deeply regrets any omission and pledges to correct errors called to its attention in subsequent editions.

Unless otherwise acknowledged, all photographs are the property of Scott Foresman, a division of Pearson Education.

Photo locators denoted as follows: Top (T), Center (C), Bottom (B), Left (L), Right (R), Background (Bkgd)

Opener: Corbis; 1 Getty Images; 3 (Inset) Corbis, (B) Getty Images; 4 (T) ©Royalty-Free/Corbis, (Inset) ©Comstock Inc.; 5 (BL) Getty Images, (BR) ©Comstock Inc.; 6 Index Stock Imagery; 7 Corbis; 9 Paul Barton/Corbis; 11 Getty Images; 13 (BL) Corbis, (BR) Roger Ressmeyer/Corbis; 14 Getty Images; 15 PhotoEdit; 20 Sion Touhiq/Corbis; 22 Getty Images

ISBN: 0-328-13549-6

Copyright © Pearson Education, Inc.

All Rights Reserved. Printed in the United States of America. This publication is protected by Copyright, and permission should be obtained from the publisher prior to any prohibited reproduction, storage in a retrieval system, or transmission in any form by any means, electronic, mechanical, photocopying, recording, or likewise. For information regarding permission(s), write to: Permissions Department, Scott Foresman, 1900 East Lake Avenue, Glenview, Illinois 60025.

8 9 10 V0G1 14 13 12 11 10 09 08

What Was It Like Back Then?

It's easy to take our society's dazzling array of modern technology for granted. In your lifetime, incredibly sophisticated electronic devices such as DVD players, cell phones, and CD players have all become standard household items. So too have systems like the **Internet,** that amazing network of internationally linked computers. Can you think of a time in your life when you didn't have such high-tech products at your fingertips? You probably can't! But for a long time, items such as these, which people use for various forms of learning, entertainment, and communications, were only available to the government or to scientists working at universities and research institutions. And it wasn't so long ago that these technologies didn't exist at all.

To help you see how modern technology has changed our lives over the past few years, you'll soon read about a fictional high-school student from 1975 named Sally. By examining Sally's routine, you'll be able to see how people lived just a few decades ago, before they had access to the modern technology that we now enjoy.

Millions of Americans own cell phones, computers, DVD players, and GPS systems. But as recently as twenty years ago these items were not available to the public.

Sets of encyclopedias take up lots of shelf space, but the information they contain can now be stored on one CD-ROM disk.

Doing Research

Imagine that you've been whisked back to 1975, and are now watching Sally as she does research at the library for her social studies report on Lewis and Clark. Sally wishes her parents would buy her a set of encyclopedias to help with school reports, but they've told her that encyclopedias cost too much, and would take up too much room on the bookshelf. Sally would rather be at the town park, playing softball with her friends, but the library isn't open for much longer and will be closed tomorrow. With the report due in three days, Sally has run out of both options and time.

After several hours at the library, Sally returns home, where she is delighted to see that a letter has come from her older brother Alex, who's spending a year in Europe. Alex's correspondences usually take two weeks to arrive, but Sally scrutinizes the envelope and sees that this time it only took nine days for her brother's letter to reach her.

Sally reads her brother's letter, engrossed by his descriptions of European castles. Alex apologizes for not sending any pictures, explaining that he didn't have time to get the photos developed. But he points out that it would've been futile to get them developed in the first place, since he couldn't afford the postage to send them!

Writing, Shopping, and Entertainment

Sally decides to write back to her brother that night. She prefers using her father's typewriter, but makes too many mistakes typing with it, and when she attempts to fix her errors the correction fluid only seems to make them more obvious! Frustrated by this dilemma, Sally resigns herself to writing by hand. Recalling the embarrassment she felt in the past when her brother pointed out her mistakes, Sally remembers to haul out her parents' big, heavy dictionary in order to confirm the spelling and usage of certain words she has difficulty with.

In her letter Sally describes the shopping trip that she and Mom went on during the previous weekend. All they were looking for was a blouse to complement one of Sally's skirts, but they ended up going to four different clothing stores before finding something that they both found suitable. It had taken an endless amount of searching through the racks, and Sally had nearly given up in fatigue.

With some typewriters, corrections can only be made by applying correction fluid directly to the paper.

The memory of the frustrating shopping trip lingered in Sally's mind. She wrote to Alex, "Why couldn't there be a service that listed which blouses were at what stores? That way, people would know in advance which store to go to, and never have to spend hours going from place to place in search of the things they wanted!"

The thought of the shopping trip made Sally recollect the time a couple of weeks ago when she had stopped at a bookstore with Mom to see if a book they had ordered was in yet, only to have the clerk inform them that the book still hadn't arrived. "It was just like what happened when we went shopping for the blouse!" Sally wrote. "If someone invented a service that tells a customer when a product they've ordered has arrived, life would be so much easier!"

Right when Sally finished her letter, she got a phone call from her friend Deb. There was a new movie out that they both wanted to see, and they agreed that when tomorrow's paper came, they'd check it for the movie's times and locations. When Sally hung up, she thought, *I wish there was something I could check right now to find out those movie times. Then Deb and I wouldn't have to wait for the paper!*

Sally sighed. She knew how lucky she was to have the things that she had. But would it ever be nice if some things were more convenient!

> Rotary phones, which required the caller to spin a dial, were used until recently.

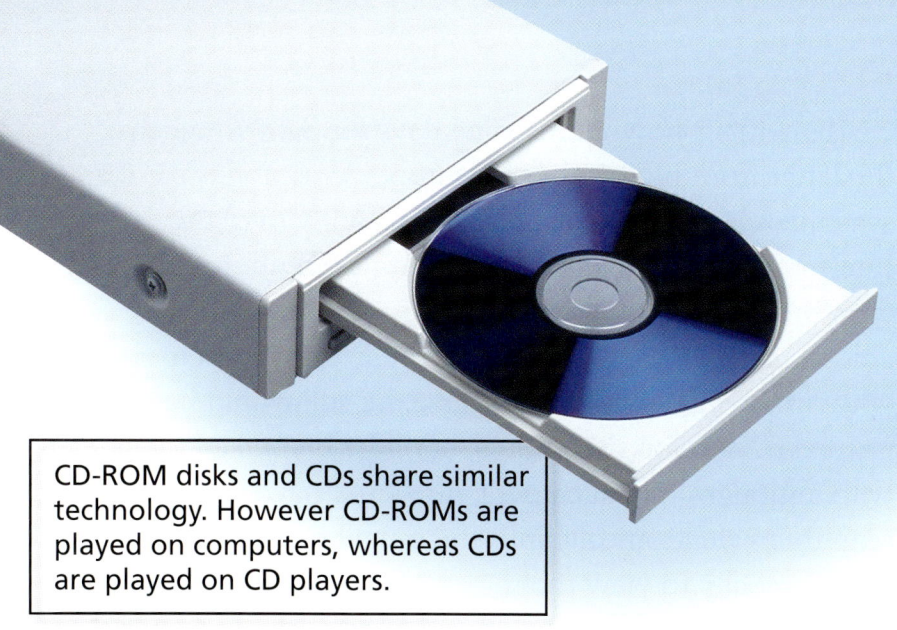

CD-ROM disks and CDs share similar technology. However CD-ROMs are played on computers, whereas CDs are played on CD players.

What Has Changed?

Having read Sally's story, surely you noticed how even the simplest activities were much more time-consuming back in 1975. For starters, in order to do research for her social studies report on Lewis and Clark, Sally had to go to the library. If Sally were doing her report today and had a home computer, she could do research from home on the Internet. By doing her research online, Sally would have been able to save an incredible amount of time!

Remember how Sally wished that her parents would buy her a set of encyclopedias? They had pointed out that it would cost too much and take up too much space. But now there are encyclopedia sets that fit into the palm of your hand! They are written onto **CD-ROM** disks, which utilize the same digital technology as regular CDs and are identical to them in size and shape, but are read by a computer's CD-ROM drive as opposed to being played on a CD player. CD-ROM disks take up a miniscule amount of space, are more durable than the delicate pages of a book, and are much more affordable than a set of encyclopedias.

Writing Letters

Remember how it took nine days for Sally to receive the letter from her brother in Europe? With today's e-mail systems, letters can be exchanged instantly over the Internet. E-mail also allows people to send photos along with their letters as attached files. So if Alex had been able to use it, he could have sent the extra photos to Sally without paying postage. Of course, he still would have needed to develop the photos, but if he had taken them with a modern digital camera, he would have been able to scan them electronically into the e-mail he sent to his sister, allowing Sally to view them.

Now think back to the letter that Sally wrote to her brother. Today, computers come with programs called **word processors,** which have eliminated the need for typewriters. Word processors allow the writer to make corrections to text as it's being written, making correction fluid unnecessary. They also come with devices that check spelling. These devices, which function like dictionaries, are similar to CD-ROM disks in that they store massive amounts of information that would take up thousands of pages if written out. If Sally had been able to use a spell-checker, she wouldn't have needed a dictionary to find out how to spell a word. Her spell-checker would have let her know when she'd misspelled a word, and it would have given her the correct spelling.

Buying Things

Remember how frustrating Sally found it to have to go to four different clothing stores to find the blouse she wanted? If she were shopping for the blouse today, she could find out which stores carried it by checking out the stores' Internet Web sites. Such a search would have taken her minutes, as opposed to the hours it took to go to the actual stores. Moreover, if Sally didn't want to pick up the blouse, she could order it from the store's Web site and have it shipped to her home. She would even be able to track the blouse's exact location as it was being shipped by entering its tracking number into the shipping company's Web site!

Today's technology also could have helped Sally and her mother when they went to see if the book had arrived. Using a computer, Sally's mother could buy the book from the bookstore's Web site and request that an e-mail be sent notifying her when it came in. Then she could pick it up whenever she wanted.

Internet Web sites allow people to shop online, without having to leave home.

Planning Entertainment

Finally, think back to when Sally talked on the phone with her friend Deb. Remember how annoyed she was because she had to wait until morning to find out when the new movie was playing? Because of that, Sally and Deb weren't able to make plans that night over the phone.

Nowadays, if Sally and Deb had been planning to see a movie, they wouldn't have had to wait until the movie times were published in the next day's paper. Instead, they could have used the Internet and checked Web sites that display movie times and locations. Or, they could have called a phone-based movie information hotline, which would also provide movie times and locations.

Today you can find out at any time when and where movies are playing by checking Internet Web sites.

The Computer Age

Whether you're checking movie times, looking for blouses, doing library research, or engaging in any other activity, the Internet is an incredible tool for accessing information. But it wasn't just the Internet that was unavailable to Sally in 1975. At that time, people had to make do without digital photography, word processing, spell-checking, cell phones, CD-ROM disks, DVDs, and much more.

Early computers took up vast amounts of space, yet could only perform a tiny fraction of the tasks done by today's computers.

These items are recent technological innovations, each of which has affected our lives in different ways. Their combined effect has been described as a technological revolution, a revolution known as the **Computer Age.**

The Computer Age has transformed our lives in many ways, making it as influential as the **Industrial Revolution** of the 1800s. The Industrial Revolution caused tremendous changes by introducing steam power, railroads, cotton gins, and similar breakthroughs into American society.

Although we call it the Computer Age, many of the technologies we use were invented long before today's computers. Cars, TVs, telephones, radios, alarm clocks, and toasters (just to name a few) all predate the computer. People have made improvements to these items since they were invented, but they are still used in the same way that people used them a half-century ago.

In comparison, as computers have evolved, people have come up with more and more uses for them. Many of these uses have been thought up only in the past decade—in your lifetime!

The Rise of Personal Computers

Computers were first developed around the time of World War II, but they didn't become part of everyday life until the early 1990s. Such a time lag between when something is invented and when the average person starts to use it happens frequently, because new technology needs to be *refined*. Computers have been refined continually since World War II, as smaller, less expensive, more powerful, more versatile, and easier-to-use models have come on the market. And the market for computers grows daily, as more people learn how to use them through school and work.

Although personal computers first became available to the general public during the late 1970s, they were of limited use for people who weren't computer specialists, and they were too expensive for most people to buy. But as more software was developed and more refinements were made, computers became less costly and more useful.

One of the biggest breakthroughs in personal computing came with the invention of word processing programs, which allow their users to write out, store, and print written materials. Word processing programs also allow their users to correct mistakes, make instantaneous text changes, run spell-checking, and save texts for future modifications. They have made computers practical. However, another invention has eclipsed them in popularity: the **World Wide Web.**

Many people view the World Wide Web as a recent phenomenon, but ideas relating to it have been circulating for over a half-century. In 1945, a scientist named Vannevar Bush wrote an essay describing how it would be possible to create an electronic system that both created and navigated through links to documents stored on microfiche, which is a special kind of film. However, the World Wide Web as most people know it really began in 1989, when Tim Berners-Lee proposed that it be built.

The World Wide Web is an information system that allows people to review and retrieve Web sites and pages found on the Internet. Before the Web was widely available most people viewed personal computers as luxuries, but afterwards they were seen as necessities! By the mid-1990s, demand for computers had increased substantially, due to the ease with which regular people could access the Internet. The Internet and World Wide Web allowed people to do things that had not been possible before. They even created a new leisure activity—"surfing" the Internet.

The first personal computers came out three decades ago. It would take them many years to gain widespread popularity.

The Internet's Many Uses

Of all the Internet-based technologies, **e-mail** (an abbreviation for "electronic mail") may have changed our society the most. With e-mail, you can exchange documents in a matter of mere milliseconds with people on the other side of the world, as long as you send your e-mail to an active e-mail account and use the proper address.

Because of e-mail, many workers are now just as productive at home as at the office. This has led to a rise in **telecommuting**, which is when a person does all of his or her work from home using either a personal computer or a similarly sophisticated, high-tech electronic device.

Before telecommuting, there were plenty of people who worked from home, but the number of at-home workers skyrocketed in the 1990s in response to the convenience created by e-mail and the World Wide Web. In 2001, it was estimated that thirty-two million people, or approximately 24 percent of the workforce, telecommuted. Roughly nineteen million of those workers were using Internet-based services in order to do their work.

Nowadays many workers take advantage of their personal computers to telecommute.

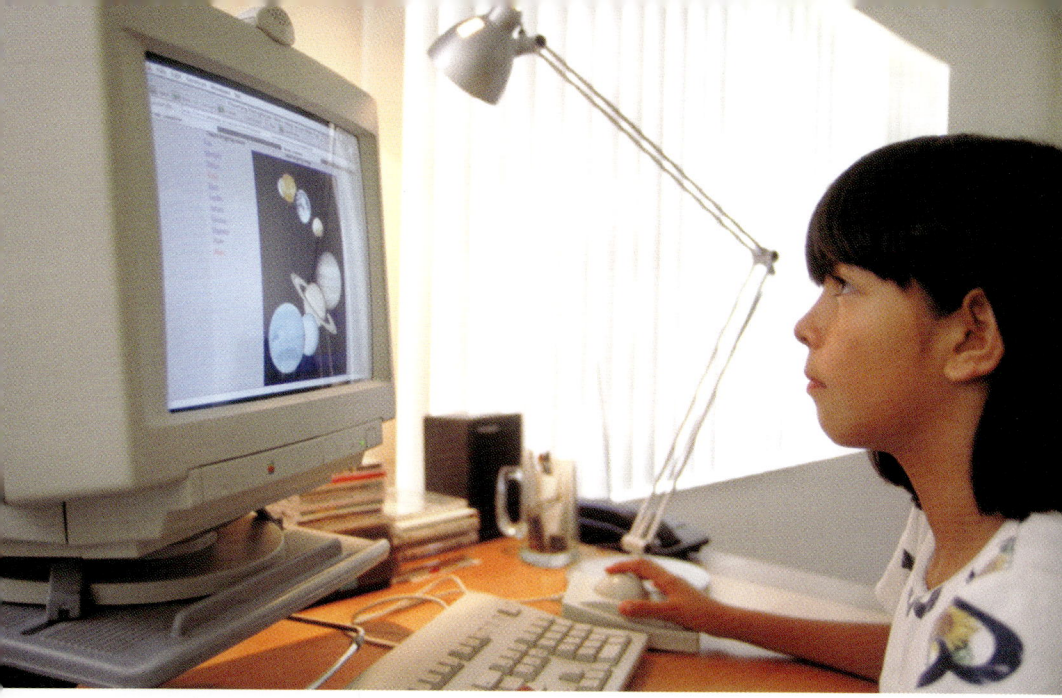

Computer-based "distance learning" has grown rapidly over the past few years, as more and more people embrace its convenience.

Computer-based "distance learning" (online classes) is another activity made possible by the Internet. In the past, when everything was sent by regular mail, distance learning involved weeks of waiting. Students and teachers had to wait for homework, quizzes, exams, and other course-related materials to be sent through the mail.

Now, with computer-based distance learning, neither students nor teachers have to wait like that anymore. The materials necessary to teach and learn a distance-learning class can be downloaded or sent by e-mail. The ease with which distance learning courses are conducted has fueled an explosive growth in the number of online universities.

Computer-based distance learning has become incredibly popular in recent years. But other Internet-based services have also impacted learning. One such service is the **search engine.**

Search engines first appeared in the 1990s. They allow people to access the data stored on the Internet's Web sites.

How do search engines work? You go to the search engine's Web site and type in your search terms. The search engine takes those search terms and locates every Web site it can find that contains them. While the search engine is scouring the World Wide Web for relevant Web sites, it analyzes the sequence of your search terms. Based on this analysis, the search engine ranks the Web sites that it finds. Finally (although it sounds as if this process takes a while, in fact it's executed instantaneously), the search engine comes back with a listing of Web sites, ranked from most to least important based on the search terms used.

Search engines can also translate text from one language to another. This has created great opportunities for scientists, scholars, and researchers, as they are no longer confined to working with materials written only in their language. As you can see, search engines do much more than merely hunt down information!

Search engines allow people to track down information on the Internet with incredible speed and efficiency.

The Quickening Pace of Change

One of the interesting things about the ongoing technological revolution is that it has happened so fast. Having lived your entire life during the Computer Age, you might not be able to perceive the breathtaking rate at which change has happened. But compared with past technological revolutions, the Computer Age has progressed at the speed of light!

One of the main reasons for the Computer Age's rapid pace of change is that modern scientists and inventors have been able to take advantage of so many previous breakthroughs. When most people hear the word "computer," they think of a sophisticated electronic device that comes equipped with a variety of high-tech accessories, such as e-mail systems, Web browsers, drives that play CD-ROM disks, software programs that share and download digital files, and systems that allow users to create digital films, music, and more. But computer-like machines, such as calculators, have been around for centuries. So too has the mathematical system of calculus, which helps explain the science behind computers.

The following two-page time line describes just some of the events that have been associated with the Computer Age. Some of the facts you may have already known about. Others will be brand-new to you!

Milestones of the Computer Age

1984
- The CD player, which uses computer-based digital technology, is invented, along with CDs.

1985
- Microsoft Windows 1.0, a software program designed for personal computers, is first released. The Windows program would come to dominate the world of personal computers.
- The CD-ROM, which allows CDs to be played on computers, is invented.

1986
- Internet connections are established to link universities and research centers to the National Science Foundation.

1987
- The one-millionth Apple Macintosh computer is sold.

1988
- The first "worm," or destructive, computer-based "virus," appears on the Internet, raising concerns about computer crime and security.
- The first transatlantic fiber-optic cable, linking American computers to European ones, is set in place.

1989
- A major milestone is passed as more than 100,000 computers are now linked to the Internet.
- The Internet's World Wide Web system, or WWW, which is still in place today, is devised.
- The first commercial Internet Service Provider, or ISP, is founded. The system, which uses regular phone lines, makes the Internet available to the general public.

1990
- Computer engineers are now able to design computer chips containing one million separate parts.

1992
- The spread of computer viruses through the network of computers linked to the Internet encourages the development of antivirus software.

1993
- Web browser systems are developed, allowing people to navigate the Internet's pages with ease.
- The White House sets up its own Web site, to be accessed through the Internet by regular users.

1994
- Search engines first become popular on the Internet, allowing people to track down a wide variety of online information based on the search terms they input.

1995
- "Streaming" technology, which allows people to watch and listen to stored computer files as video and audio clips, gains popularity on the Internet.

1998
- Congress passes the Digital Millennium Copyright Act. The Act prevents people from using computer software to steal or reproduce copyrighted materials.

2000
- Demonstrating the Internet's phenomenal growth, there are now an estimated one billion pages available to be viewed on the World Wide Web.
- Internet Service Providers (known popularly by the acronym "ISPs") change their method of transmitting online data by switching from phone lines to cable lines.

2004
- The U.S. Department of Energy announces that it will build a "science research supercomputer" capable of processing 50 *trillion* calculations per second.

Costs and Benefits: Computer Viruses

When you saw the term "virus" on the time line, what images came to mind? Did you picture little "bugs" traveling across computer networks, "infecting" everything they touch?

In truth, **computer viruses** are bits of computer code designed by people. They cause damage to computer systems and data. People spend billions of dollars each year fixing the damage caused by computer viruses.

Computer viruses are very harmful, but they've also played a major role in motivating people to improve the security of computer systems. This sort of mixed impact is typical of the Computer Age. We can't say that computer viruses have been purely negative, because they have inspired computer scientists to invent important protections for computers. But obviously we can't say they've been positive, given the damage they cause. In the end, computer viruses, for better and worse, are a reality that people will have to deal with for as long as there are computers.

Computing Our Progress

As computer viruses show, the Computer Age has had an uneven impact. It has produced some great benefits, some of which we explored earlier by comparing Sally's lifestyle in 1975 with today's computer-enhanced lifestyle. At the same time, it has created many new problems that few people could have foreseen in 1975.

We have already talked about the costs associated with computer viruses. We can also consider the costs of the Computer Age by reviewing one of Sally's experiences from a different perspective. If Sally buys a blouse from an Internet Web site and has it shipped to her, she won't know what it feels like before it arrives in the mail. By buying it directly from the store, she has the opportunity to try it on beforehand. Sometimes, "low-tech" is better, because it forces people to have hands-on interaction with things that computers can't describe, like a blouse's actual feel.

Despite these problems, almost everyone agrees that life has been made more convenient by computers. Information that once would have taken weeks, months, or years to collect is now literally "at our fingertips." Through computers and similar technological advances, we can carry out tasks and activities that were unthinkable only three decades ago. For all of these reasons and more, you are lucky to be living in the Computer Age!

◀ Computer viruses prevent computers from carrying out their instructions. They were given their name because they "infect" computers that are linked together in a network.

Now Try This

Taking Stock of Today's Technology

You may not think about it much, but as a student growing up in the twenty-first century, you're both highly knowledgeable and heavily dependent on all sorts of modern technologies. Whether it's the DVDs you watch on a DVD player, the CDs you listen to on a CD player, the cell phone you use to talk to your friends, or any other kind of modern device, you're surrounded by a variety of machines, systems, and products that your parents and grandparents didn't have access to when they were growing up.

The following activity asks you to imagine what it would be like if you had to live for a period of time without modern technology. Read the steps carefully before you begin and have your teacher answer any questions you might have. Then get ready to project yourself into the time before the Computer Age!

Here's How to Do It!

- On a separate piece of paper, describe in your own words how you use modern technology over the course of a typical day. Mention the technology that you use at home, school, the library, and friends' houses.
- Now imagine that you are going to spend a few months camping in a remote area. The modern technology that you wrote about earlier has to be left at home. The campsite where you're staying has electricity, and there is a small town nearby, but you won't have the opportunity to use CD players, DVDs, cell phones, GPS systems, or any similar form of modern technology.
- Next, using the same paper, describe how you would spend your free time, now that you've been denied access to modern technology. How will you communicate with your family? What sacrifices or substitutes will you have to make to compensate for the technology-dependent activities that you will be missing out on?
- Finally, get together in small groups with your classmates. Have each person describe the technology they use today, and how they'd attempt to live without it. Select a recorder for the group and have that person take notes on each person's response. If there's time, have your group present what you have talked about.

Glossary

CD-ROM *n.* a compact disk that plays on a computer's CD-ROM drive.

Computer Age *n.* the time period in which computers have transformed modern life.

computer viruses *n.* programs, designed by people, that do damage to computers or data.

e-mail *n.* system of sending messages using computer technology.

Industrial Revolution *n.* the changes in technology of the 1800s that affected the way people lived.

Internet *n.* a worldwide computer network, linked by telephone lines or cables, that is used to send messages, data, and other services.

search engine *n.* a program that helps people find information on the Internet.

telecommuting *v.* working from home using a personal computer.

word processors *n.* computer programs that edit, store, and retrieve documents and texts.

World Wide Web *n.* an information system that allows people to review and retrieve Web sites found on the Internet.